Good For Me
Water

Sally Hewitt

WAYLAND

Notes for Teachers and Parents

Good for Me is a series of books that looks at ways of helping children to develop a positive approach to eating and drinking. You can use the books to help children make healthy choices about what they eat and drink as an important part of a healthy lifestyle.

Look for water when you go shopping.
- Look for water in different forms at your local supermarket.
- Find and identify foods that contain high amounts of water.
- See how many different ways you can have water during the day.

Talk about why water is important to us and how it plays a vital part in a balanced diet.
- Water is a liquid. Discuss what foods contain water.
- Talk about the ways water helps to keep us strong and healthy.

Talk about how we feel when we are healthy and the things we can do to help us to stay healthy.
- Eat food that is good for us.
- Drink plenty of water.
- Enjoy fresh air and exercise.
- Sleep well.

First published in 2007 by Wayland
This paperback edition published in 2008 by Wayland
Copyright © Wayland 2007
Wayland
338 Euston Road
London NW1 3BH

Wayland Australia
Hachette Children's Books
Level 17/207 Kent Street
Sydney NSW 2000

Produced by Tall Tree Ltd
Editor: Jon Richards
Designer: Ben Ruocco
Consultant: Sally Peters

British Library Cataloguing in Publication Data
Hewitt, Sally, 1949-
Water. – (Good for me!)
1. Water – Physiological effect – Juvenile literature
2. Health – Juvenile literature
I. Title
641.2

ISBN: 9780750256223

Printed in China
Wayland is a division of Hachette Children's Books.
www.hachettelivre.co.uk

Picture credits:
Cover top Alamy/SuperStock, cover bottom dreamstime.com,
1 dreamstime.com, 4 Alamy/David R. Frazier Photolibrary, Inc.,
5 Alamy/Photo Network, 6 Dreamstime.com/Álvaro Germán, 7 Tall Tree Ltd,
8 Alamy/David R. Frazier Photolibrary, Inc., 9 Dreamstime.com,
10 Dreamstime.com, 11 Dreamstime.com/Suprijono Suharjoto,
12 Dreamstime.com/Chris Schlosser, 13 Dreamstime.com/Steve Lutes,
14 Dreamstime.com/George Bailey, 15 Dreamstime.com/Joan Ramon
Mendo Escoda, 16 Alamy/Howard Harrison, 17 Dreamstime.com/Johnny
Lye, 18 Dreamstime.com/Martin Green, 19 Alamy/Dynamic Graphics
Group/IT Stock, 20 top left Dreamstime.com/Cristian Ardelean, top right
Dreamstime.com/Rafa Irusta, bottom left Dreamstime.com/Verena Matthew,
bottom right Dreamstime.com, 21 top right Dreamstime.com, top middle
Dreamstime.com/Alvin Teo, top left Dreamstime.com, bottom left
Dreamstime.com/Stephen Walls, bottom right Dreamstime.com/Ian Francis,
23 Corbis/Tom Stewart

Contents

Good for me

We need to eat food and to drink water to live and be healthy. Water falls as rain. Rain is collected in lakes called **reservoirs**.

Water comes to our homes through pipes and out of taps.

Water is important because we drink it and use it to prepare and cook meals. We need to drink plenty of water to stay healthy. Water is good for you!

Water helps our bodies to turn food into energy so that we can exercise and have fun.

Water for life

All living things, including people, need water to live and grow. If we do not have enough water, our bodies start to dry out and we feel very thirsty.

This plant is dead because it has not been given enough water.

Water is so important to us that without it we quickly become ill. If we do not drink any water for a while, we might feel tired or get a headache.

About two-thirds of your body is made up of water.

Lunch box

Take a bottle of water to drink instead of fizzy drinks in your lunch box.

Using water

Your body is losing water all the time.
There is water in the air you breathe
out and you lose water when you go
to the toilet.

You can see the water in your breath
when you breathe out on a cold day.

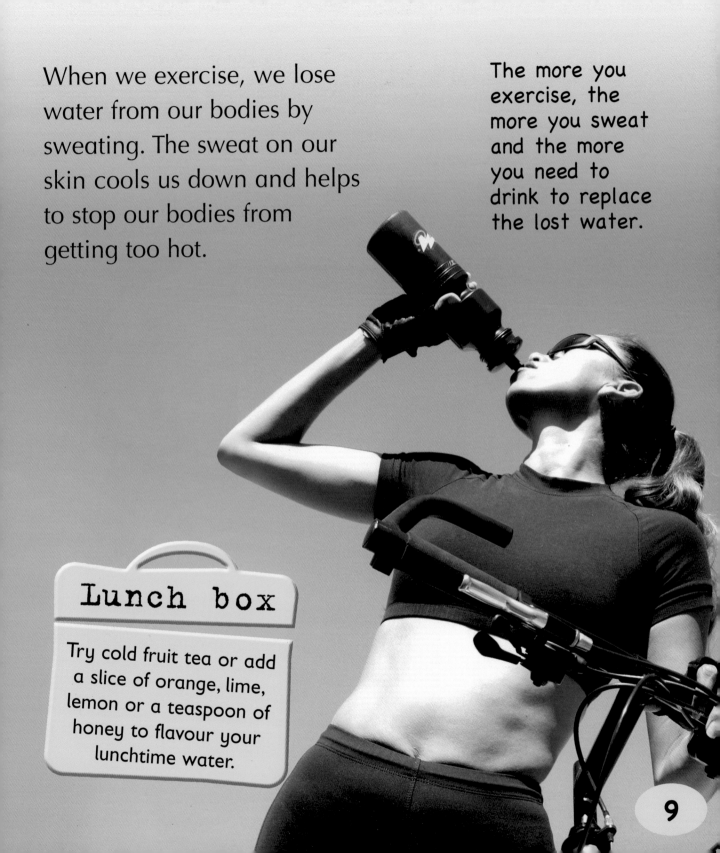

When we exercise, we lose water from our bodies by sweating. The sweat on our skin cools us down and helps to stop our bodies from getting too hot.

The more you exercise, the more you sweat and the more you need to drink to replace the lost water.

Lunch box

Try cold fruit tea or add a slice of orange, lime, lemon or a teaspoon of honey to flavour your lunchtime water.

Replacing water

You should have drinks throughout the day to replace the water your body has lost. Do not wait until you are thirsty before having a drink of water.

You can drink from water fountains while you are out and about.

Not all of the water you need comes from drinking. The food you eat also contains water and this can replace some of the water that your body has used.

Lunch box

Watermelons are fruit that store water. Add a slice of watermelon to your lunch box.

Try to drink about 2 litres, or eight glasses of water every day. The rest of the water your body needs will come from the food you eat.

The water cycle

The air around us is full of water. The water collects together to form clouds. Sometimes, the water in clouds falls to the ground as rain.

Water that falls on the ground trickles into streams and rivers.

Streams and rivers flow into lakes and the sea where the water **evaporates**. That means it goes up into the air in tiny droplets. The droplets of water collect together to form clouds again. This is called the **water cycle**.

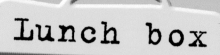

Lunch box

Put some chilled water in a thermos flask to keep it cool until lunchtime.

Rain falls from the clouds and the water cycle starts all over again.

Clean water

Water is not always clean. Rain falls through dusty, smoky air. There is waste and oil in rivers and in the sea, as well as dirty water from our homes.

Water is kept in large reservoirs before it is treated and cleaned at a **water plant**.

Dirty water is full of germs that can make us ill. Water needs to be treated at a water plant to kill germs and to make it safe to drink.

After it has been treated at a water plant, the water is sent down pipes and into our homes.

Lunch box

Ask an adult to make a drink with hot water. Keep it warm in a thermos flask for lunch on a cold day.

Buying and storing water

You can buy bottled water in shops. Carry a small bottle around with you to drink from throughout the day. You can refill the bottle and use it again. This will help to reduce plastic waste.

Shops sell still, sparkling, plain and flavoured bottles of water.

If it is kept in a very cold place, water will **freeze** to form a solid called ice. When it is warmed, ice melts back into liquid water.

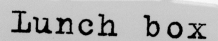

Lunch box

Half fill a plastic bottle with water and freeze it. It will stay cold on a hot day.

You can freeze water in cubes. Add the ice cubes to make ice-cold drinks.

Water and food

Water is used in cooking. Vegetables, fish and fruit are **boiled** or **steamed** in water to cook them.

Potatoes can be cooked by putting them into boiling water.

Some foods contain lots of water, such as **citrus fruits** and leafy vegetables. Water is also added to recipes to make liquid meals, such as soups and stews.

Lunch box

Ask an adult to help you cook sliced apples in water and honey. Eat them cold for lunch.

As well as lots of water, citrus fruits, such as oranges, contain **vitamin** C which is important for healthy gums.

19

Food chart

This page shows dishes that are made using water. Have you tried any of these?

Soup

Rice salad

Cooked mussels

Pasta

This page shows different foods that contain lots of water. Have you tried any of these before?

Tomato

Lettuce

Cucumber

Citrus
fruits

Watermelon

A balanced diet

Water is an important part of a balanced diet. This chart shows you how much you can eat of each food group. The larger the area on the chart, the more of that food group you can eat. For example, you can eat a lot of fruit and vegetables, but only a little oil and sweets.

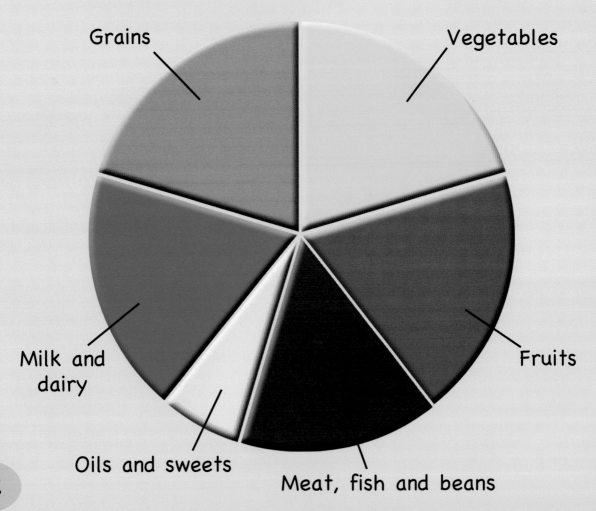

Grains

Vegetables

Milk and dairy

Fruits

Oils and sweets

Meat, fish and beans

Our bodies also need exercise to stay healthy. You should spend at least 20 minutes exercising every day so that your body stays fit and healthy.

Always make sure there is an adult nearby when exercising in water.

Glossary

Boiled Food that has been cooked in boiling water.

Citrus fruits A kind of fruit with thick skins and juicy insides, such as oranges and limes.

Evaporates When a liquid changes into a vapour.

Freeze When a liquid turns into a solid.

Reservoir A large lake where water is stored before it is treated at a water plant and made safe to drink.

Steamed Food that has been cooked in the steam from boiling water.

Vitamin Important substances found in food. For example, vitamin D helps you to grow strong bones.

Water cycle How water moves around the Earth. It falls as rain, flows into rivers, lakes and the sea, evaporates into the sky, where it forms clouds and falls as rain again.

Water plant A place where water is treated and made safe to drink.

Index